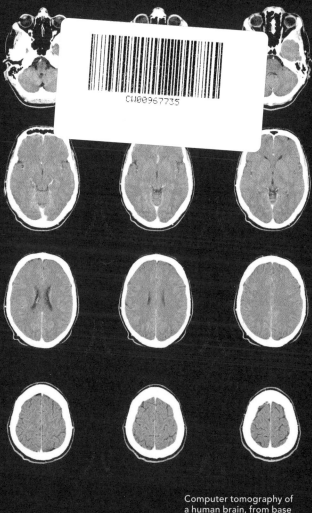

Computer tomography of a human brain, from base of the skull to top, after a 20-minute episode of loss of the left visual field. No physical reason for the loss of vision was found by the scan.

Series 117

This is a Ladybird Expert book, one of a series of titles for an adult readership. Written by some of the leading lights and outstanding communicators in their fields and published by one of the most trusted and well-loved names in books, the Ladybird Expert series provides clear, accessible and authoritative introductions, informed by expert opinion, to key subjects drawn from science, history and culture.

Every effort has been made to ensure images are correctly attributed; however, if any omission or error has been made please notify the Publisher for correction in future editions.

MICHAEL JOSEPH

UK | USA | Canada | Ireland | Australia
India | New Zealand | South Africa

Michael Joseph is part of the Penguin Random House group of companies whose addresses can be found at global.penguinrandomhouse.com

First published 2018
001

Text copyright © Hannah Critchlow, 2018

All images copyright © Ladybird Books Ltd, 2018

The moral right of the author has been asserted

Printed in Italy by L.E.G.O. S.p.A.

A CIP catalogue record for this book is available from the British Library

ISBN: 978–0–718–18911–2

www.greenpenguin.co.uk

Penguin Random House is committed to a sustainable future for our business, our readers and our planet. This book is made from Forest Stewardship Council® certified paper.

Consciousness

Hannah Critchlow

with illustrations by
Stephen Player

Ladybird Books Ltd, London

What Exactly is Consciousness?

We are all unique. Each person on this planet holds their own, highly individual sense of reality shaped by their distinct set of past memories. As a result, we all experience the world, and respond to it, in our own way. This individualized experience underpins the essence of consciousness: to be aware of our surroundings, hold a subjective view of the world, and then interact with the environment with our own perspective.

How such consciousness arises has been debated for centuries. Traditionally, this was the domain of philosophers, working mainly in the realm of the theoretical. Recent technological advances have, however, made it possible to produce objective measures of consciousness, to visualize its development and to observe what happens as it fails; scientists have joined in the study.

As we understand more precisely how the brain operates, it would seem inevitable that we should get closer to a straightforward explanation of consciousness. Yet, paradoxically, for many this knowledge increases the allure and mystique of the phenomenon, challenging our current definition of consciousness as perhaps too simplistic. Our increased knowledge also produces pressing ethical considerations – are other animals, or even plants, conscious? Can we create conscious robots? If so, should they have the same legal rights as humans? Can we enhance our own levels of consciousness? Is it possible to invoke a shared collective consciousness? And as we understand more about the nature of consciousness, what are the ramifications for the concept of free will? Perhaps even our ability to ponder and debate these issues forms a part of what it means to be conscious . . .

Where Does Consciousness Live?

As they tried to understand how consciousness forms, philosophers were first fascinated by the question of where in the body it arose.

In the fourth century BC, Aristotle reasoned that since the heart is vital for life, consciousness must originate there. Around 500 years later Galen, the Greek physician, surgeon and philosopher, moved upwards. He conducted numerous dissections and, as his anatomical knowledge grew, so did his appreciation for the brain. Galen believed that consciousness lived in the clear, colourless liquid that surrounds and cushions this organ. Galen called it the 'psychic pneuma', a magical fluid that acted as the breath of life and instrument for the soul.

In the sixteenth century AD, Descartes continued with Galen's hypothesis that consciousness arose in the brain, but focused its source on the tiny pineal gland located at the brain's centre. He believed this structure opened when you looked up so that animal spirits of memory were permitted entry. He reasoned that looking to the ground closed off these memory spirits, thereby allowing deep conscious thought. Descartes argued that the physical world (the chair you are sitting on, the apple you are eating) is entirely separate to the non-physical world of your individual perceptions and sensations (the feeling of the cold, hard seat on your legs, the satisfying crunch of the apple as you take your first bite, the fresh smell and sweet taste that you experience). He proposed that the pineal gland allowed communication between these two worlds to enable you to experience sensations and form your unique view of the world. This theory is known as Cartesian dualism – reflecting the idea that the body and the mind occupy separate realms.

What is It Like to be a Bat?

In order to probe the nature of consciousness, philosophers would conjure up thought experiments: using imagined scenarios to uncover the nature of reality.

One of the most famous thought experiments on consciousness is 'What is it like to be a bat?', debated by the philosopher Thomas Nagel in 1974. He argued that since consciousness is the ability to form a subjective view of the world, it is useful to imagine being something. For example, if you want to know if a table is conscious, then ask yourself, what is it like to be a table? Most people would concur that a table cannot feel or think, that it is simply a physical item and so does not hold a unique view of the world. A table is just unconsciously there.

But what about a bat? Can you imagine what it might be like to fly around the night sky, relying on echoes to navigate, eating bugs, then sleeping upside down? What would it be like to be a bat? If you can start to answer this, then bats are conscious, Nagel argued. What about a bamboo plant? A fruit fly? A bumble bee? A cucumber? Or a computer? Can you imagine being these? If so, perhaps they are also conscious.

Nagel took this thought experiment a step further by suggesting that if you were able to metamorphose into a bat's body, you would still not be a true bat because you were not born a bat. You would retain your human perspective and therefore could never possess a bona fide bat mindset. You would be more like a manbat, with a bat body and human mind.

The Philosophical Zombie

A second hotly debated thought experiment explores the idea of a philosophical zombie.

Imagine a zombie, one that looks identical to you. To the outside world, it is completely indistinguishable from you. It can receive information and respond to it, in the same way as you can, but it is a zombie. For example, if this zombie stubbed its toe, it would behave exactly as you would (exclaim 'Ouch!' and move away). However, this zombie could never experience the sensation of pain, as you and I can, since it is simply a zombie.

Many people believe that such a zombie could never logically exist. Therefore discussing its existence is pointless. Others counter that since we can imagine such a zombie, this in itself provides a vital clue to understanding consciousness.

There are around 1 million people on this planet who, like the philosophical zombie, cannot experience the feeling of pain. Some of these people have a mutation in the gene called PRDM12 which switches off their pain receptors at birth. All of these unfortunate patients cannot perceive pain, and so they do not learn to respond to the sensation by removing themselves from its source or by shouting out for help. As a result, although these individuals are unanimously agreed to be conscious while alive, they are commonly covered in scratches and bruises, and often die at an early age from serious accident.

These pain-free, short-lived individuals help us to appreciate that the concept of a philosophical zombie ignores an important aspect underlying consciousness: that sensations help us to build a perception of the world, allowing us to learn from the environment in order to survive.

How Does the Brain Operate?

It is now generally accepted that consciousness arises in the brain from interaction with the outside world. Happily, over the previous few decades there has been a technological revolution, enabling us to peer into the brain and observe how it operates – with high resolution and in real time – in fully conscious, moving, learning, mammals. Such a toolkit is useful for understanding how our perception of the world is formed and why we think, and behave, as we do. Your brain contains around 86 billion nerve cells. To scale it down to something more palatable, if we took a dot of brain tissue the size of a sugar grain, it would contain roughly 10,000 nerve cells. Even more incredible: each of these nerve cells connects to around 10,000 others, forming the most densely packed and complicated circuit board imaginable.

The brain can be described as a circuit board since it employs the power of electricity to send signals via its nerve cells, which are connected with each other in neural networks. When an electrical signal reaches the end of one nerve cell, a chemical called a neurotransmitter is released, crosses the gap between the cells (a 'synapse') and activates the next cell to continue the signal across the circuit board. A thought is, essentially, a fizz of electricity. There are also inhibitory nerve cells, which act as stop signals at traffic lights, to prevent further activity across particular junctions.

This system underpins our thoughts and dictates how we interact with the world. Yet it is now possible to achieve magnificent feats tweaking it. We can identify and locate the discrete pathways that are responsible for specific behaviours, such as addiction or depression. Implants can be surgically attached to these circuits buried deep in a patient's brain to switch off symptoms simply using the power of electricity.

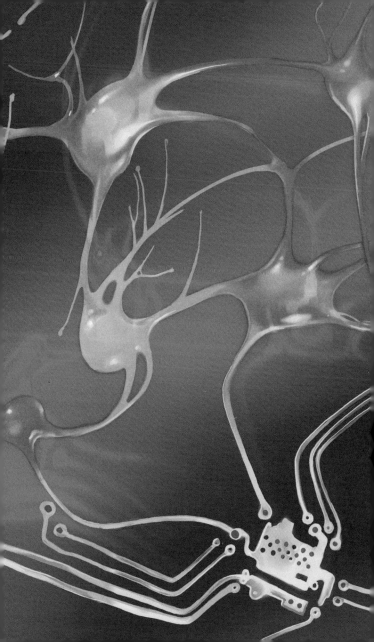

Brain Input and Output

In addition to the complexity of the brain's circuit board there are also approximately 10 billion nerve cells in the rest of the body. These operate in a similar way as those in the brain, using electrical signals. A stimulus from the outside world may activate these cells in your body to respond quickly and simply in reflex, for example, instructing you to drop a hot plate even before you have realized it is burning you. Alternatively, the electric signal is conveyed to your brain, where the information is processed and integrated with prior knowledge, and explicit instructions are generated on how to respond.

Returning to the case of a stubbed toe: pain receptors are activated and send an electric signal rapidly to the brain's thalamus. This region relays the signal to the sensory cortex, where it is interpreted as a sharp pain. A signal is initiated in the motor cortex region of the brain, directing signals to your mouth to exclaim 'Ouch!' to alert others to your pain and your need for assistance. Further signals are sent to your leg, instructing you to relax the muscles and sit down in the nearest comfortable seat to recover. Meanwhile, slower impulses travel through the neighbouring nerves to form a throbbing ache felt through the entire toe, a warning to treat the area gingerly while it heals.

This process demonstrates that regions across the brain work together to form a perception of an experience and to direct a response. These regions interact as a complex network, much like airport connections between major cities. The study of brain networks is called 'connectomics' and is key to understanding how consciousness occurs.

Is Consciousness the Ability to Learn and Remember?

In order to form a subjective view of the world, and to survive in it, we draw on our earlier experiences. As we learn, new connections form between nerve cells. As the new skill is practised, or the learning relived, the connections strengthen, so the learning is consolidated into a memory. If the memory is repeatedly visited, it becomes the default route for electrical signals in the brain and in this way learned behaviour becomes habit.

The majority of connections between nerve cells occur on minuscule structures termed 'dendritic spines'. They have the ability to change shape in response to electrical activity. As learning occurs, thin, wormlike structures, the precursors of dendritic spines, reach out to make contact with the neighbouring active nerve cell. As the electric signal continues to travel between the cells, proteins that help process the signal are recruited. The dendritic spine swells and a bulbous mushroom-like head is formed. If the connection is repeatedly excited, it will split into two daughter spines, thereby doubling the circuit connections.

The majority of learning is initially laid down in the hippocampus, a small seahorse-shaped structure buried deep in the centre of your brain. Simultaneously, a trace of the new learning forms in the prefrontal cortex, the region behind your forehead, where it is thought to be consolidated into a stored memory. Connections between these regions occur in addition to other areas of the brain, allowing new incoming information to integrate with memories and form your perception of the world. These findings pose the question: is consciousness simply our ability to learn and remember?

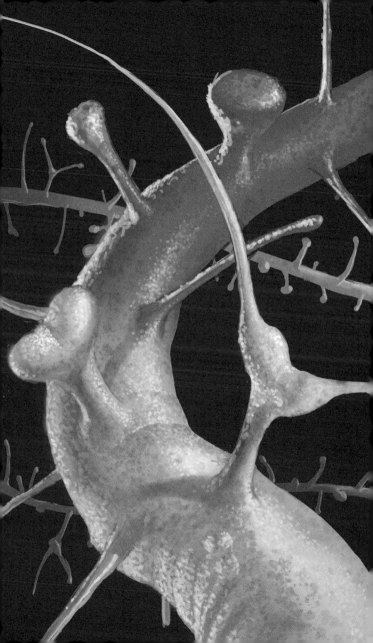

Is Consciousness the Ability to Assume?

In addition to learning for survival, the brain uses prior experience to filter information, create shortcuts in processing and enable rapid analysis to form your view of the world. Opposite are two photos of a mask. The back end of the mask is hollow, but you perceive it as a face, even though the shadows are telling you this is not the case. This bias for seeing faces is built from experience and is so strong that it causes us to ignore the information the shadows are giving us. The hollow-mask illusion demonstrates how your brain employs prior experience to make assumptions. Although we are constantly bombarded with information from the world, we can rapidly form a perception of it. In this way, your individual sense of reality is based on a culmination of your unique set of past experiences.

People with schizophrenia are immune to this illusion. They perceive the photo in its literal sense – as the back end of a hollow mask. (NB: it does not mean you are schizophrenic if you do not see two faces; psychiatric diagnosis is not that simple.) People with schizophrenia collect the same incoming information through the eyes, but the 'top-down' process of interpreting and making assumptions is altered. Analysis of their brain anatomy and activity shows there are fewer connections in the hippocampus and prefrontal cortex. As a result, their brains do not as readily filter information based on past experiences. This may explain why people with schizophrenia experience distortions to reality, seeing and hearing things that other people do not, and may experience problems with reasoning, planning or flexibility in thought. These results suggest that consciousness evolved to help us survive, so that we can learn from, and interact with, the world.

Perception Without Conscious Awareness

In 1974 psychologist Lawrence Weiskrantz documented Patient DB: a man who was adamant he was completely blind on one side of his visual field. Yet when asked to detect, localize and discriminate among visual stimuli presented there, he achieved a much higher accuracy than would be expected from chance alone. DB was exhibiting perception without conscious awareness, or 'blindsight'.

Analysis of his brain and those of subsequent patients similarly diagnosed with the disorder revealed that they all suffered from damage to the primary visual cortex, a region located at the back of the brain. In the succeeding years, numerous studies have been conducted to measure brain activity in both human patients and animals modelling the condition. The results suggest that conscious awareness relies on networks of electrical activity across the whole brain, with the primary visual cortex operating as a hub to help orchestrate the visual conscious broadcast. When the hub is damaged, perception can occur without awareness.

Picking apart this phenomenon provides clues about the power of the unconscious mind. For example, a patient with blindsight is able to navigate their way past objects they are not aware of. This shows that awareness isn't the whole story: although consciously realizing something seems to help us react to an ever-changing world, often the brain has actually already made a decision for us, without our knowing.

Blindsight also forces us to consider the possibility of consciousness existing in several forms, since patients only have impaired visual consciousness whilst the other facets of their behaviour are intact.

The Smell of Fear

Emotions help to drive our basic behaviour. They can be interpreted as the end result of information-processing without our conscious awareness. For example, happiness occurs when incoming signals activate the reward pathways in our brains, the chemical dopamine is released and the nucleus accumbens region is electrically stimulated. This feeling is so pleasant that we are motivated to seek the positive experience again. Eating, exercise and sex activate these brain circuits, helping us to survive and reproduce as a species. Addictive drugs operate by hijacking this system.

At the other end of the spectrum are feelings of fear, which help to protect us from danger. Activation of the amygdala, an almond-shaped structure in the middle of the brain, directs a hormonal cascade, including the release of adrenaline, that prepares our bodies for fight or flight. It also causes the release of olfactory chemical signals in our sweat to alert those around us, making them more alert to potential danger. At the same time, we may lay down memories that connect the amygdala to other areas of the brain – these associations help to protect us if a similar situation arises again.

The way that our emotions have been wired into us can be taken advantage of. Film composers, for example, can manipulate our emotional responses to great effect. The *Jaws* theme creates a sinister feeling of suspense with its chilling, minor crescendos and jerky dissonant chords. Such non-linear sounds are similar to a baby's scream and trigger a biologically ingrained response to feel fear by making us think our young are threatened. Horror-film music composers also employ low-frequency sound waves that emulate the effect of adrenaline, activating nerves in the stomach to produce that sensation of dread in the pit of your belly.

Why Do We Sleep?

Are there different levels of consciousness? Can we measure it on a scale as we do for height or weight? It was previously thought that consciousness decreased as one fell asleep, reaching an unconscious level during deep sleep. It is, however, now generally accepted that your brain is highly active during sleep, and again the capacity for learning and memory is key.

The sleeping brain consolidates learning from the previous day into memories, strengthening the connections in the circuit board, so that we wake to see each new day with a fresh perspective. This helps to explain why chronic sleep disturbance is associated with cognitive decline, such as dementia and Alzheimer's, and is linked to illnesses characterized by an altered reality, such as schizophrenia.

Studies looking at the waves of electrical activity in the brain during sleep help us to understand how sleep supports these phenomena. Small, flat metal discs (electrodes) are attached to the scalp to pick up the brainwaves passing under the skull via an electroencephalogram (EEG). The electrical activity across the brain oscillates with frequencies between 0 and 50 Hz (up to 50 cycles per second). These cycles of oscillations have been categorized into six bands, each helping to support a specific aspect of consciousness, and the results illustrate some interesting differences between levels of brain activity while awake and asleep.

What Happens in Your Brain as You Wake and Sleep?

Mu waves (8–12 Hz) in the brain are involved in directing muscles in our bodies so that we can respond to the world through movement and speech. Alpha waves (8–15 Hz) are associated with calm thoughts and creativity, beta waves (16–31 Hz) are associated with attention, concentration and focus. The fastest gamma waves (> 32 Hz) are important for integrating incoming information, helping to form our perception of the world.

Fully alert and awake individuals have, as you may have guessed, more of these higher frequencies of brain electrical activity. As we start to fall asleep, these faster brain waves start to drop off, replaced by slower electrical pulses called theta waves (4–8 Hz) and delta waves (< 4 Hz). It is during deep sleep that memory consolidation occurs, with the slower electrical oscillation activity thought to support this.

The exact purpose of dreaming is not yet clear but it may possibly highlight important events yet to be consciously processed. Dreaming occurs during Rapid Eye Movement (REM) sleep, when brain activity ramps up in frequency. The waves become desynchronized, with chaotic patterns of electrical pulses and fast brain waves re-emerging to resemble the EEG brain profile of wakefulness. During lucid dreaming (when the sleeper is aware that they are dreaming), the EEG profile resembles that of wakefulness even more closely.

These studies indicate that differing patterns of electrical activity across our brain may enable the various facets of behaviour that make up consciousness.

Altered Consciousness: Lessons from Vegetative-state Patients

Vegetative state is a disorder of consciousness where an individual is unaware of, and unresponsive to, their surroundings. If this state lasts longer than a few weeks, it is referred to as a persistent (or 'continuing') vegetative state, with traditionally no hope for recovery.

It has recently been discovered that approximately 20 per cent of such patients are actually fully aware but simply unable to control their movements to interact with the environment. It is now possible to communicate with these individuals by reading their brain activity. Patients lie in a brain scanner and are asked a series of questions. Their responses are measured by quantifying the amount of oxygen delivered to particular brain regions. This oxygen supports the high-energy demands of the brain's electrical activity and so an increase in the amount of oxygen is a marker for thought.

The patient is first asked simple yes/no questions, such as 'Is China in Asia?' They imagine playing tennis if the answer is 'yes' or walking down a corridor if the answer is 'no'. Different areas of the brain light up depending on which activity is being thought of. The patient is then trained to imagine different activities for different durations and in this way all of the letters in the alphabet are coded for and the patient can communicate. 'Reading' the brain like this can help to identify which patients might make a recovery.

Scientists have built on this to identify further hidden signatures. Instead of using brain imaging (which is expensive and time-consuming), it is possible to measure electrical activity in the brain using an EEG.

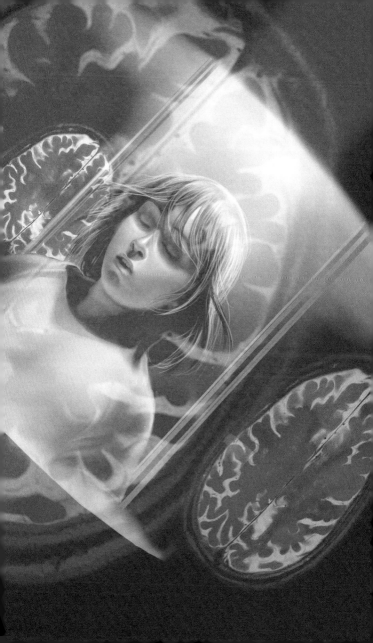

Can We Increase Consciousness?

Electric signals rush around your brains at approximately 120 miles per hour. Is it possible for all of your brain's 86 billion nerve cells to be electrically active at the same time? With high frequencies of oscillations? Would this increase your level of consciousness? Or would it cause you to blow a fuse?

Recent blockbuster films such as *Limitless* and *Lucy* have imagined how pharmacological manipulation could increase consciousness. This concept has been science fiction fodder for decades, and an important driver provoking social debate. Such 'smart drugs' are, however, no longer simply a figment of writers' imaginations. Modafinil and methylphenidate are commonly prescribed to help treat sleep disorders or attention deficit hyperactivity disorder (ADHD), to boost alertness, attention and focus. They are also sometimes used by students during exam revision and a growing number of academics admit to taking them to help with arduous committee meetings or writing grant applications, as there is some evidence that they increase working memory.

Rather than opening up the mind, as depicted in the films, these drugs seem to do the opposite, to narrow down the circuit board's activity so that attention does not wander. These drugs localize brain activity to discrete areas so that the individuals can focus on a specific task, helping to filter out competing and distracting incoming sensory signals. They also seem to increase the speed of electrical activity in the brain, increasing alertness. Some hypothesize this will come with a cost to creativity, and the long-term side effects of these drugs in healthy individuals are not yet known.

Altered Consciousness:
Lessons from Creativity and LSD

Psychedelics, such as LSD, are famous as being the trippy drug of the 1960s and '70s, when they were believed to increase consciousness, dissolving your ego so that you felt at one with nature. They are illegal, and it has been difficult to find funding for studies investigating how they operate, particularly after the public uproar when it was discovered that the CIA had surreptitiously tested LSD on unsuspecting human subjects, hoping to use it as a tool for mind control.

There has, however, been a recent resurgence of interest in LSD. Coders in Silicon Valley report to be illegally microdosing LSD to help boost their creativity and problem-solving skills, while preliminary clinical trials indicate that low doses of the chemical might be useful in treating post-traumatic stress disorders, addiction and depression.

In a study, two groups of volunteers were injected with either a placebo (just salt water) or microdoses of LSD. The individuals were unaware which group they were assigned to. Their brains were scanned, to analyse levels of electrical activity as inferred by oxygen concentrations. The brains of those taking LSD lit up like a Christmas tree.

LSD appears to open up the circuits in the brain, decreasing the filtering of information. The brain profile is comparable to a naive childlike networking brain, in which the more restricted thinking we develop from infancy to adulthood is reversed. It seems to dissolve the brain's preference to make assumptions and hold preconceptions based on prior learned behaviour. This explains how it might be helpful in treating certain psychiatric disorders, by breaking particular negative habits of thinking and opening up alternatives.

Lessons from Meditation

Derived from the Latin *meditatio*, meaning 'to think, contemplate and devise', the art of mediation has been practised since antiquity. Proponents claim it can improve self-awareness (the ability to introspect and distinguish oneself from the environment), increase concentration and improve the control of conscious thought. Through extensive practice, it is claimed, one can transcend consciousness, go beyond all thoughts, sensations and perceptions, to experience a pure state where you are aware only of consciousness itself.

Research on the topic is in its infancy; however, a number of studies have examined the brains of long-term practising Buddhist monks and volunteers as they undergo a meditation course. The results indicate that meditation engages clusters of activity in the brain including the caudate region (thought to have a role in focusing attention), the hippocampus (involved in learning and memory and conscious control of mind-wandering) and the medial prefrontal cortex (involved in self-awareness). It is hypothesized that over time meditation activates these brain regions to induce new nerve cells to be born in the hippocampus, increases connectivity between the regions and promotes insulating fat to wrap itself around the brain cells, protecting electrical signals. Another theory is that the practice promotes the health of existing nerve cells by regulating the immune system, and reduces stress, thereby decreasing the damaging effects of the stress hormone cortisol on the brain.

Oscillations of electrical activity across the brain also change during meditation, with increased alpha and theta waves. This may signify a state of relaxed alertness conducive to mental health.

Are Other Mammals Conscious?

Traditionally, other animals have been viewed as simple automatons. Recent research has, however, blown away such presumption. Other mammals may not be able to report their perceptual experiences in a language that we can understand but they can learn, remember, plan for the future, experience emotions, solve problems and communicate with each other.

In light of recent findings, the 2012 Cambridge Declaration on Consciousness publicly proclaimed: '. . . the weight of evidence indicates that humans are not unique in possessing the neurological substrates that generate consciousness. Non-human animals, including all mammals and birds, and many other creatures, including octopuses, also possess these . . .'

Indeed, the architecture of the mammalian brain and its mode of operation is so similar between species that rodents are commonly employed to model aspects of complex human psychiatric conditions including schizophrenia, depression, anxiety, autism, learning difficulties and dementia, in order to understand how the conditions arise and how we can better treat them.

Animals even appear to dream while asleep; for example, sleeping dogs can be seen twitching their legs and then sleepwalking to chase imaginary objects. Apes, monkeys, elephants and dolphins all exhibit self-awareness – they can recognize themselves in mirrors. While dogs may not understand the concept of mirrors, they are able to conceive self, being attracted to sniffing other dogs' excrement rather than their own. Perhaps canines are letting us know that the key to consciousness is being able to avoid your own crap?

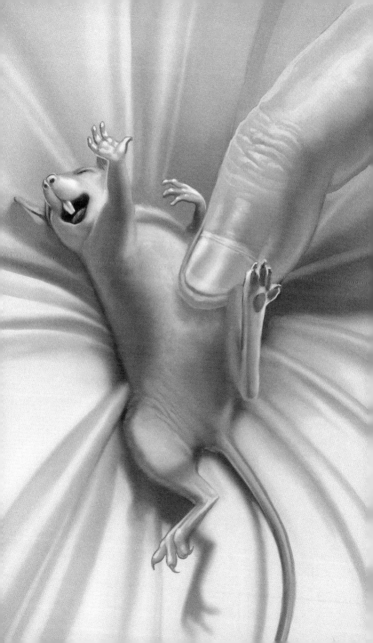

Are Birds Conscious?

In ancient Greece, 2,500 years ago, the famous storyteller Aesop described how on a hot day a crow, desperately thirsty, finds a pitcher. Unfortunately, there is only a small amount of water sitting at the bottom. The intelligent bird drops stones into the container, raising the water level, little by little, until it is high enough for the crow to drink.

Today, crows in captivity display a similar problem-solving skill, even though they have never observed others behaving in this way. When a similar test is given to four-year-old children, the crows beat them: the children simply stare frustratedly at the beaker.

Indeed, crows have been described as the ultimate problem-solvers; they can successfully complete complex puzzles in sequence to retrieve food rewards. In the wild, crows can be seen creatively adapting to city life, for example, dropping encased nuts from the sky on to the roads below next to traffic lights and then waiting for the passing cars to break the hard shells with their tyres to release the nuts inside. The birds then use their beaks to press the pedestrian crossing button, bringing the traffic to a halt so that they can safely retrieve their food.

Interacting with the environment creatively, and adapting to it, in order to ensure survival is thought to be an important aspect of conscious behaviour, demonstrating awareness and responsiveness. Should birds also be considered as conscious? If so, perhaps the term 'birdbrain' should no longer be a derogatory one.

Are Insects Conscious?

Insects exhibit incredibly complex behaviours, especially in communities. Ants, for example, build transport systems and communicate food routes via scent trails, and they even farm together, inserting fruit seeds into cracks in tree bark, fertilizing it with their faeces and at harvest-time saving the fruit's seeds for future crops. In fact, evidence suggests they started planting fruit crops long before human agriculture evolved.

The honeybee is another fascinating insect. It employs the 'waggle dance', an incredible figure-eight dance, to communicate with other members of the colony, sharing information about the direction of, and distance to, new patches of flowers yielding nectar and pollen. In spring, when bees swarm, they collectively settle at a new hive within a couple of days of starting their house search. Indeed, ants and bees are an interesting subject of study in the field of collective consciousness – showing how a group can unite in their perceptions for social goals.

Such capacity to learn from the environment, plan for the future, work together and creatively solve problems means that these insects could also be viewed as conscious organisms. Insect brains do operate in a similar way to ours, albeit on a smaller scale. The honeybee brain has fewer than a million neurons, an ant brain just a quarter of this. Such relative simplicity makes their brains a lot easier to study. The more we observe and understand the behaviour of other species, the more our view on consciousness may have to shift to accommodate the realization that humans may not be, after all, really that special.

Are Plants Conscious?

Unlike other animals, birds and insects, plants do not have a centralized brain. But perhaps that is sensible – it would be a shame to break off half your mental muscle during a gale or to have it chewed away by a rabbit. Instead, plants spread their processing power across their entire body: from their roots to the tips of their leaves. Although plants may not have the same type of nerve cells as we do, they similarly use the power of electricity to transfer information around their bodies, while the genes that direct this activity are evolutionary analogues to ours. Plants even use the same type of neurotransmitter chemical messengers as those found in our brains.

Just like humans, plants can sense what is happening around them, process these signals and respond to the environment accordingly. For example, if a caterpillar is nibbling on a leaf, the plant will start to produce a chemical to repel the insect. Even if the plant is simply played an audio file of a munching caterpillar, it will respond, indicating that plants can hear.

Plants also have a sense of community. For example, if an African acacia plant is being eaten by a giraffe, it warns its neighbours by releasing ethylene gas. Downwind acacias detect this and manufacture poison that makes their leaves less attractive to the giraffe. Such community spirit is not limited to acacias: in forests across the world, trees share water and nutrients across species using a highly complex and interconnected root and underground fungal system.

It is breathtaking to realize that even a brainless plant is capable of such complex cognitive skills.

> The common garden pea, *Pisum sativum*, can be taught to associate an air current with light, and will grow through a maze towards a fan in search of a solar-power source in order to photosynthesize. The plant is making choices based on predictions informed by its prior experience.

Can Robots be Conscious?

We live in an age of artificial intelligence (AI): driverless cars on the streets, Siri on our phones, Carebots looking after our elderly. These systems employ machine learning: they take in information from the environment, use it to build a framework of reality. Inspired by the biological neural networks found in the human brain, AI systems are being developed that are able to process and integrate information rapidly and even demonstrate creativity – an important facet of conscious behaviour. Indeed, software has successfully been created that can compete with humans at writing literary prose and composing original music scores. Robots have been developed with software that recognizes emotion with a higher degree of accuracy and sensitivity than some humans. Can robots feel emotions? Well, they respond to reward; for example, Facebook developed AI that was motivated to lie during negotiations in order to get what it wanted. Robots may also be able to elicit emotions from us, with a third of humans thinking it possible to fall in love with a robot, and some commentators predicting that marriage between humans and robots will be legal by 2050.

Interestingly, robots also demonstrate a degree of self-awareness. There are robots that can recognize themselves in a mirror, for example, or employ powers of deduction to determine how they differ from other robots. Although self-awareness is a separate behaviour from consciousness (the ability to form a unique perception of the world), it is interesting that a robot can have this capability. We can now artificially create consciousness (or at least produce machines that simulate certain characteristics of it). But how good does a simulacrum have to be before we concede that it is consciousness?

Does Free Will Exist?

Do we really have the capacity for freedom of choice? Or are the decisions we make on a daily basis actually just inevitable computations? Is free will simply an illusion? A growing body of research seems to indicate we are all machines, that consciousness is the result of processing incoming signals to provoke inescapable actions.

Way back in 1985, neuroscientist Benjamin Libet devised an experiment that attempted to determine whether conscious decisions are made before or after our brains give instructions. He asked subjects to flex their wrists repeatedly at times of their choosing. He measured the movement of the wrist and the activity of the brain's motor cortex, and compared these with the time the individual reported consciously deciding to move. Precise times for the wrist movements were obtained by picking up the electrical activity of the muscles via an EEG. Similarly, electrodes were placed on the scalp to pick up electrical activity in the motor cortex with high sensitivity. Libet discovered that the brain's instruction to act came first, the conscious decision to act came 350 milliseconds later, then there was a 200-millisecond delay before the actual movement. In effect, conscious awareness occurred after the brain directed the action. This rather simple experiment has been repeated and refined since and, taken together with recent discoveries about the exact mechanisms by which our brains direct behaviour, raises fascinating and difficult questions about free will. What would happen if we all lost faith in our freedom of action? A slew of recent research suggests that eroding an individual's belief in free will fosters increasingly self-centred and impulsive behaviours whereby social rules are broken. Perhaps an illusion of free will is necessary for a smooth-running society.

What is Consciousness For?

We each have a unique perception of the world. Every person on this planet holds their own, highly individual circuit board, a brain wiring system shaped by the culmination of their past, personalized, set of experiences and memories. It is this cartography in the mind that forms our subjective view of the world, dictates how information about our world is processed, filtered and made sense of, before instructing us how to act.

This navigation system for comprehending the world is not static: new connections between brain cells form as we learn; new pathways of communication are forged as we think. This ability, this consciousness, that lives in our brain, is essential for our survival. Such a flexible framework of perception allows us to rapidly alter how we interact with the world as it changes around us.

Despite this magnificent architecture of the brain – its intricacy, dynamism and phenomenal processing power – occasionally mistakes are made. None of us are perfect. A system, however, appears to have evolved alongside this to compensate: social cohesion.

Communication provides a bridge between the faults inherent in the individual and culture; it helps a collective consciousness and civilization emerge. It transpires that in discussing our subjective view of the world with each other, we are generally more likely to arrive at a more accurate representation of the world. Therefore, the old adage that two heads are better than one appears to be true. Perhaps such a mechanism provides the only way we, as a society, can be strengthened and, as a species, we can survive to flourish.